Pedro Ienne Fernandes
Claudia Tania Pincinin

3D Printing for Dental Prosthetics

Pedro Ienne Fernandes
Claudia Tania Pincinin

3D Printing for Dental Prosthetics

And the impact on quality of life

ScienciaScripts

Cover image: www.ingimage.com

This book is a translation from the original published under ISBN 978-613-9-60203-2.

Publisher:
Sciencia Scripts
is a trademark of
Dodo Books Indian Ocean Ltd., member of the OmniScriptum S.R.L Publishing group
str. A.Russo 15, of. 61, Chisinau-2068, Republic of Moldova Europe
Printed at: see last page
ISBN: 978-620-4-10313-6

SUMMARY

INTRODUCTION

As the population ages the possibility of total or partial loss of their teeth increases, especially for those who do not have access to periodic dental treatment. Tooth loss is a dental mutilation that leaves the individual predisposed to a disease state, since this loss signals physical, biological and psychological changes. Thus, individuals who have lost their dentition feel vulnerable and limited (MENDONÇA, 2001).

According to Santos (2014), both the low socioeconomic level and the insufficient amount of basic information on prevention of dental diseases are aggravating when the subject is oral health. Coser (2005), cites as an example, that children up to five to six years of age still do not have enough fine coordination to take care of brushing alone, therefore, parents are responsible for this guidance and it is their duty to teach and supervise their children to brush their teeth correctly and frequently.

Pinto (1997, 2000) and Guimarães and Marcos (1996), indicate that edentulism (tooth loss) in Brazil has social causes and that recoverable teeth are extracted due to economic problems. The mass extraction of teeth in Brazil starts around 30 years of age (PINTO, 1997). According to Silva, Magalhães and Ferreira (2010), in Brazil, the Unified Health System (SUS) does not have an adequate structure to absorb the demand for oral health care of the population, especially in advanced age groups. This is one of the reasons why teeth that could be recovered are extracted, since this alternative is considered the fastest and most economical.

Social conditions and the practice of deleterious dentistry (extraction of dental elements as a solution for pain relief) in populations with low socioeconomic status represent a significant number in the prevalence of tooth loss (VARGAS and PAIXÃO, 2005). According to Guimarães, Pinto, Amaral *et al.* (2017), tooth loss is one of the main problems to oral health due to its high prevalence and the functional damage it causes, constituting an important health problem.

Other factors impacted are facial appearance, speech and chewing, besides that its absence negatively influences the quality of life of people. According to Goffman (1988), when an individual has a different trait or characteristic, such as a physical deformity, this individual may be rejected, leading him/her to be disabled for social acceptance. Thus, the lack of teeth may be one of these deformities that leads the person to have social problems. Despite the recognized importance of oral health, a considerable portion of the Brazilian population does not have access to health services (Ferreira *et al.,* 2005).

A solution to the loss of teeth, depending on the patient's case, may be the manufacture of a partial or total, fixed or removable prosthesis (CALEJO, 2017). Dental surgeons usually indicate dental prostheses in order to improve the harmony of the smile, increasing self-esteem, chewing function and speech of the person, which are usually impaired by the lack of teeth.

According to Ferreira (2005), at the same time that dentures are indicated as a solution for tooth loss, those who need them end up having limited access because it is not a type of product and service frequently offered in the public health system. As an alternative to the public health service, which in its majority does not provide activities focused on dental prostheses, there are private dental clinics to perform this service.

A research was made in the internet, in the site "hospedario.com" in order to raise the prices of a Fixed Dental Prosthesis for installation in a private clinic, divided in its types:

Table 1 - Prices of a Fixed Dental Prosthesis in 2021

Type of Prosthesis	Price
Single dental implant (intraosseous screw only)	R$1.370,00
Metal crown	R$470,00
Porcelain dental crown	R$1.020,00
Temporary acrylic crown	R$120,00
Featured upper or lower total prosthesis (one unit)	R$1.600,00
Temporary partial removable prosthesis (PPR)	R$630,00
Overdenture which is the socket denture over implants	R$2.200,00
Immediate total prosthesis per unit	R$1.400,00

Source: Site: https://hospedario.com.br/quanto-custa-uma-protese-dentaria-fixa/ - **Accessed on 19 Apr 2021.**

As previously presented, the main reasons that lead to the loss of teeth are the financial conditions. Thus, with the high prices of dental prostheses, it becomes unfeasible for people from poor communities to have access to this type of solution for the loss of teeth.

One of the alternatives for the conventional manufacture of dental prostheses is the use of digital workflow, with three-dimensional printing of polymeric materials, it is possible to quickly manufacture components with high geometric complexity, according to the desired needs and specifications (MARGARIDA, 2020). Being a flexible technology, 3D printing allows implants and prostheses to be produced according to the specificities of each individual patient. According to Gorni (2011) it is estimated that the time and costs savings provided by the application of rapid prototyping techniques in the construction of models, such as the 3D printer, are around 70 to 90%.

According to Yan *et al.* (2017), 3D printing is based on the principle of layered manufacturing, in which materials are overlaid layer by layer, rapidly manufacturing components with any shape, accurately accumulating material using solid modeling according to a computer-aided design (CAD) or computed tomography (CT) model. According to Ogliari (2018), in the dental field 3D printing enables the production of dental implants, prostheses, anatomical models, restorations, surgical guides and orthodontic models and aligners. For the construction of these prostheses are necessary technologies, as for example, the scanning of images, which allows obtaining the data and information needed for its construction.

Within this reality, clinics and hospitals do not need to acquire new equipment and hire specialized professionals for the construction of 3D prostheses, because the images required for the application of these techniques can be obtained using equipment already present in the daily routine of medical care facilities, such as computed tomography, magnetic resonance imaging or intraoral *scanners* specific for this scanning, which are more recent (MARGARIDA, 2020).

In recent years, there has been a remarkable advance in the application of Rapid Prototyping systems in Dentistry in various specialties (DUTRA *et al.* , 2017). Since within these terms there is the possibility of being made prostheses via 3D printer, which can be cheaper than conventional ones, this paper seeks to identify the main types of 3D printing techniques to make dental prostheses and present what are the main limitations of 3D printing in dentistry.

For this, a research protocol was applied to search for scientific articles that dealt with the topics of 3D printing techniques related to dental prostheses and also on the limitation of 3D impressions in the dental area, seeking to contribute to the academy so that people in this area can have access to an updated material and, in addition, if this project is completed in full, it will have a great social contribution.

THEORETICAL REVIEW

Prosthodontics and 3D Printing

The main objective of Prosthodontics is to replace one or more missing teeth in the mouth of the patient, aiming to return the smile and also self-esteem. In addition to improving aesthetics, the prosthesis contributes to health, because it returns the oral functions, improving chewing, phonetics and even breathing (ZAHR, 2020). Masson (2020) defines prosthesis as a device that is implanted in the body that aims to make up for the lack of an organ or restore a lost function, in the case of teeth, the goal is to replace missing or compromised teeth.

There are several types of prostheses that vary in terms of material and technique applied, depending on the case. This variation makes the final result change in terms of comfort, stability and quality of life. Within the many different possibilities of prostheses, there are some that stand out and end up being more common, as is the case of total prosthesis (PT) (ZAHR, 2020).

The PT (Figure 1), better known as "denture", is indicated to artificially replace all the teeth of the upper and/or lower arch. This type of prosthesis is supported only by gum tissue and other anatomical structures of the mouth and is removable. According to Zahr (2020), the total prosthesis is the best known solution among Brazilians, being produced usually in acrylic resin. In the application of this technique it is essential that the dentist and prosthodontist are very detailed to bring more reality to the patient's smile and respect their natural features.

The partial removable prosthesis, also called "*Roach*", is indicated for patients who still have teeth in the mouth and it works through a metal structure with the presence of clips, responsible for being retained in the remaining teeth promoting retention and stability when chewing, besides protecting them. Its composition is usually metallic, of an alloy of nickel-chromium or chrome-cobalt. There is a variant of this type of prosthesis known as Partial Removable Prosthesis Flex, which is produced with flexible resin, which removes the need for the use of clips and structures present in traditional prostheses. The trend is that the result has a more natural appearance (KAIRALLA, 2018).

The Fixed Prosthesis is indicated to replace large missing spaces (2 or more teeth) and can be produced both in porcelain, being more expensive and resistant, or in acrylic resin or composite. Its application is fixed and when it is applied for prosthesis of only one tooth it is known as Crown. Because it is a type of fixed prosthesis is necessary the implant, making the cost higher, so that the preference of Brazilians is the use of removable prostheses (MASSON, 2020).

An important concept to be highlighted is that of the difference between dental implant and prosthesis, which according to Zahr (2020) is that the implant is a metal structure surgically introduced in the jaw region, made to replace the roots that have been lost, while the prostheses, although they can also be fixed, most often are removable structures placed more superficially than the implants.

According to Motti (2018), the future of dentistry is digital, no matter if the dentist works in a small office or in a dental clinic or hospital, 3D printing brings many advantages, besides transferring the work of prosthetic laboratories to the dental clinic. Dental offices are already making good use of resin 3D printing to make faster service for patients who need dentures.

3D printing is a rapid prototyping technique that offers a wide utility due to its vast option of solutions and applications given its flexibility of use. According to Rodrigues Jr *et al.* (2018) 3D printing is an area of manufacturing engineering that is characterized by stages of construction of parts by automatic deposition, layer by layer from a virtual model controlled by computer programs. The 3D printer has allowed the concept of customization to expand in scale, and is considered a disruptive technology, i.e., a technological innovation that has spread across several industrial lines, including scientific and health research (SILVA and MAIA, 2014).

The first technique for producing a prototype from a virtual file was discovered in 1984 by Chuck Hull and ended up being called stereolithography, or in English *Stereolithography* (SLA). This rapid prototyping technique was defined by the inventor himself as a machine that follows a method of building solid objects by successively printing thin overlapping layers of a material curable by ultraviolet radiation energy (HULL, 2015).

According to Dabague (2014), stereolithography makes use of a resin sensitive to UV emission that polymerizes by means of radiation, thus, the UV rays are systematically projected on the layers so that the construction occurs vertically from the base to the top (*bottom-up*). Besides the possibility of alternating the angle of the source itself, there are SLA 3D printers that use a mirror to direct the UV light while its emission source remains static. For ease of understanding, an image will be presented to explain Stereolithography (SLA).

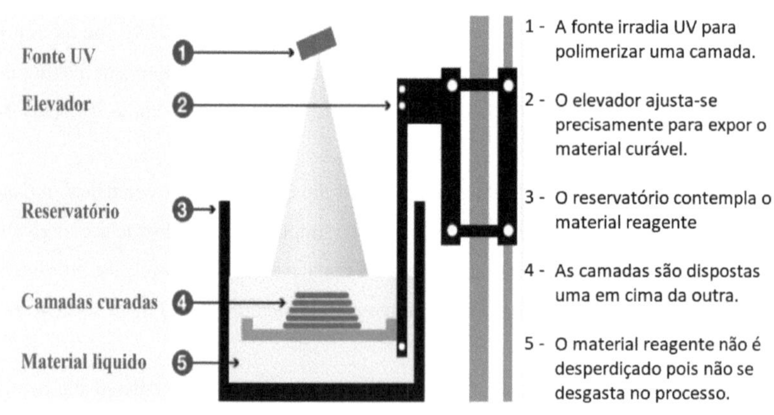

Figure 1: 3D Printing Technique: SLA
Source: DABAGUE (2014).

In 1989 Scott Crump developed another 3D printing technique, which became known as *Fused Deposition Modeling* (FDM). Then other 3D printing techniques, such as the technique of laser *sintering*, or *Selective Laser Sintering* (SLS) in English, the technique of laser melting, or *Selective Laser Melting (*SLM) in English, and also *Polyjet*, but none of these techniques were so widespread and used as the first two (DABAGUE, 2014).

Despite the differences between the techniques, all have the similarity of the final construction being performed from digital files, making the costs of production, storage, transport, among others are reduced (CASEY, 2009). According to Dabague (2014), the comparison made between the different techniques turn to the fundamental aspects of 3D printing, such as: available materials, accuracy, finish, resistance and availability bichromatic or polychromatic. In the following table it is possible to compare the different 3D printing technologies and their respective materials of use:

Table 2 - 3D printing technology and its Use Material

Technique	Material
SLA - Stereolithography	Light-curing polymers and resins
FDM - Fused Filament Deposition	Thermoplastics, eutectic blends and metals
SLS - Laser Sintering	Metal alloys: titanium, aluminium, stainless steel
SLM - Laser Milling	Thermoplastics, metal powders and ceramics
Polyjet	Combined resins and photopolymers

Source: Adapted from OLIVAREZ (2010).

Being a technique of additive manufacturing, such technique presents the advantage of the possibility of printing very complex parts using the right amount of raw material, avoiding the waste generated by subtractive manufacturing (milling). Besides this, 3D printing presents as main advantage the possibility of generating parts quickly, with quality and at a low production cost. Before the existence of additive manufacturing techniques, the production of a prototype was considered as laborious, since its development depended a priori on obtaining a mold, and then a model that now is no longer necessary due to 3D printing, which performs the construction of prototypes from files that are digitally modeled.

METHODOLOGY

According to Pagani (2018), an important feature of scientific research is that it should be replicable so that any researcher who tries to follow the methodology presented can reach the same conclusions as the author about a particular study. Given that the volume of information in the area of scientific research grows increasingly, decision making becomes more complex, and for this reason it becomes essential to seek support in tools that assist in this process.

For these reasons, the present work will be guided by a research methodology known as *Methodi Ordinatio,* the methodology uses pre-established criteria to create a process of qualification of the articles obtained to ensure a qualified systematic literature review base. This is a specific methodology of systematic literature review that locates, selects and evaluates existing studies so that it is possible to perform an analysis and synthesis of information, given that the evidence is reported in a way that allows clear conclusions to be reached about what is and what is not known (PAGANI, 2018).

Finally, *Methodi Ordinatio* is a multicriteria decision making methodology (MCDA) used in the selection of articles to compose a bibliographic portfolio. There are three analysis criteria for a scientific article: the number of citations, the impact factor and the year of publication. Based on these three criteria, the entire methodology is broken down into nine different stages.

Step number one is the establishment of the research intention, which will indicate which is the line of research that the work will be linked. The research intention turns out to be one of the most important points of the article and usually this ends up being broader and more generic than the analysis problem itself, basically it dictates which direction the work should take so that the proposed objectives are achieved. Bringing to the reality of this article, this line of research is related to 3D printing within the dental universe.

Step two consists of conducting a preliminary search with keywords related to the search intention in the databases. Through this it is possible to define which are the best keyword combinations and which databases should be used for research, arriving then in step 3 of the methodology. Within this, research with keywords related to dentistry, dental prosthesis, 3D printing and limitations in the area were made in 4 different research bases for the tests to occur, which were: ScienceDirect, Emerald, Scopus and PubMed.

Step three consists in defining what will actually be the keywords, the databases and what will be the temporal delimitation. The search terms that best performed and were selected are: ' "3D printing" AND "dental prosthesis" ' and ' "limitation" AND "3D printing" AND "dentistry" ', the first with a greater focus on the 3D printing technique in relation to dental prostheses, and the second focused on the limitations of 3D printing in dentistry. Within this, the bases that best presented results for the search were: ScienceDirect, Emerald and Scopus, due to their volumes of publications within the keywords. Finally, it was decided to establish a temporal delimitation in the last four years (2017 to 2021) to ensure that the topics addressed were still close to the reality they are today, thus avoiding outdated information.

In step four, the objective is to conduct a definitive search in the databases in search of articles, books, theses and dissertations within the parameters previously established. The table below shows the relation between keywords and databases:

Table 1 - Raw results of the systematic search

Keywords in the Title, Abstract or in the Keywords	**Science Direct**	**Emerald**	**Scopus**	**Total**
"3D printing" AND "dental prosthesis"	8	13	43	64
"limitation" AND "3D printing" AND "dentistry"	8	4	21	33
Total	16	17	64	**97**

Source: Own Authorship Research Data (2021).

In all, 97 articles that fit within the search parameters were collected within these three search bases.

Step 5 consists in the application of filtering procedures, so that the whole process becomes as standardized as possible. Within these procedures there are three that receive greater emphasis: the first topic is the removal of duplicates through some kind of reference manager, which in the case of this article was used *Mendeley*; in the second topic we have the elimination of articles by reading their title, in which those with thematic distant from the searched is eliminated; and in the third topic we have the elimination of books, chapters and *conference papers* whose has no impact factor whatsoever, thus making the work less complex and confusing.

The idea of this methodology is to be practical and it is based on an efficient philosophy, so the following actions will be applied for filtering and elimination: within the Mendeley software a scan behind duplicates happened, in which 12 articles were removed. After this 8 articles were eliminated via their title and 20 articles were eliminated after their full reading due to escaping the research topic. Finally, 7 articles with no impact factor were eliminated. The total number of articles to compose the portfolio at this moment is 50.

For enrichment of the research and greater depth in the theme about 3D printing techniques in general, 11 articles (Chart 2) referenced within these 49 found will be used to compose the tab of results and discussions, thus, the number of articles for analysis is 61.

Chart 2 - Articles used for thematic enrichment.

Title	Authors	Year
The Application of Rapid Prototyping in Prosthodontics	Jian Sun, Fu-Qiang Zhang	2012
Porous 3D modeled scaffolds of bioactive glass and photocrosslinkable poly(ε-caprolactone) by stereolithography	Laura Elomaa, Anne Kokkari, Timo Närhi, Jukka V. Seppälä	2013
The future of dental devices is digital	Richard van Noort	2012
3D printing in dentistry	A. Dawood, B. Marti Marti, V. Sauret-Jackson	2015
3D printing with polymers: Challenges among expanding options and opportunities	Jeffrey W. Stansbury, Mike J. Idacavage	2016
Evaluation of a complete denture trial method applying rapid prototyping	Masanao Inokoshi, Manabu Kanazawa, Shunsuke Minakuchi	2011
Accuracy of computer-aided design/computer-aided	Sebastian B.M. Patzelt. Shaza Bishti, Susanne Stampf, Wael Att	2014

manufacturing-generated dental casts based on intraoral scanner data		
Rapid prototyping in dentistry: technology and application	Qingbin Liu, Ming C. Leu and Stephen M. Schmitt	2006
Selective laser sintering in biomedical engineering	Alida Mazzoli	2013
Metal Fabrication by Additive Manufacturing Using Laser and Electron Beam Melting Technologies	Lawrence E. Murr, Sara M. Gaytan, Diana A. Ramirez, Edwin Martinez, Jennifer Hernandez, Krista N. Amato, Patrick W. Shindo, Francisco R. Medina, Ryan B. Wicker	2012
Recent advances in 3D printing of biomaterials	Helena N. Chia and Benjamin M. Wu	2015

Source: Own authorship (2021).

For step 6, the identification of the impact factor, year of publication and number of citations will be performed. For this purpose, the articles were organized in a table in *Excel* software with the following columns: author, article title, year of publication, impact factor (*CiteScore* 2018) and number of citations. The author, title and year were obtained through the reference manager Mendeley and JabRef, by exporting the first to the second and then the second to the *Excel* spreadsheet. The impact factor *CiteScore* was obtained through the "A5 spreadsheet", developed by Professor Bruno Pedroso from UEPG, from the postgraduate program in health sciences and made available by the official YouTube channel of *Methodi Ordinatio*. Finally, the number of citations was manually collected by searching title by title of the articles through the *Google Scholar* search.

Step 7 consists of applying the *InOrdinatio* equation (PAGANI *et al.*, 2015) via formulas in the *Excel* spreadsheet. The formula is made up as follows:

InO = ((Impact Factor/1000) + (Alpha*(10-(AnoPesq-YearPub))) + (Citations))

The Impact Factor, the Year of Publication and the number of Citations have already been collected and Alpha is a constant that should have its value assigned from 1 to 10 according to the degree of importance of the novelty in the research area. As innovation is a very important topic in the area of 3D printing and dentistry, the value of alpha was assigned to 10. *In* this way the *InOrdinatio* value of each article can already be calculated, and for this a column with the formula presented above is added.

At the end of step 7 there should be an ordering of the articles according to the *InOrdinatio* score, in which the best placed is the one with the highest score.

Step 8 consists of locating the papers listed in full format and this can be done either directly on the journal's website or through *Google Scholar*. And step 9 consists of reading and analyzing the articles so that results, discussions and some answers are raised about the research theme. All these topics will be addressed in the following topics of the article.

RESULTS AND DISCUSSIONS

The 50 articles ranked in *Methodi Ordinatio* (Appendix 1) and the 11 articles for thematic enrichment (Table 2) were read in full and then used to compose this topic of results and discussions.

Based on the analysis of the articles it could be observed that there is a great diversity of research on the topic of dental prostheses produced by 3D printing, due to the fact that there is a high variety of possibilities of solutions for this practice.

According to Katkar, Taft and Grant (2018) in the manufacturing stage, 3D printing (additive manufacturing) is becoming a fast-growing alternative to certain processes previously performed by subtractive manufacturing. And according to Hatamleh *et al.* (2013, *apud* BARAZANCHI *et al.*, 2017) the use of additive manufacturing in oral and maxillofacial prosthetics, facial prosthesis making and cranial reconstruction has increased the use of additive manufacturing, and that when manufacturing a facial prosthesis, a degree of discomfort is associated with the use of printing material in patients to create a model of the defect site. However, facial scanning can bypass this step and directly create a 3D model of the site.

Digital workflow facilitates treatment planning procedures as well as assists direct composite restorative procedures by providing several advantages compared to conventional procedures, such as accurate translation of digital diagnostics into the patient's mouth, horizontal path of silicone index insertion, and minimized clinical intervention time (PARK *et al.* , 2020; WILLIAMS et al., 2020).

Nagarajan *et al.,* (2018) state that furthermore, using 3D printing techniques, it is possible to perform risk-free surgical training on objects that are highly mimetic of natural organs, both mechanically and architecturally. It can be used for interaction with patients to achieve better understanding and compliance for diseases that require surgical intervention.

This would lead to better surgical planning and preparation based on printing models of tumors or damaged tissues that can lead to better decision making and fewer complications.

3D printing, unlike subtractive manufacturing techniques, offers a wide variety of additive procedures allowing various raw materials to be used for the fabrication of structures. This is due to the diversity of methods used to fabricate structures using additive and layering principles. Most raw materials for additive manufacturing used for dental and medical purposes can be grouped into binder/powder material combinations including polymers (resins and thermoplastics), ceramics and metals (BARAZANCHI *et al.*, 2017).

Sun and Zhang (2012) say that for intraoral prosthodontic applications, the use of additive manufacturing techniques has several uses, including the production of impression plaster models from intraoral scans, direct fabrication of dental prostheses, and custom molds for planning or investment of definitive prostheses.

Van Noort (2012), says that the basic premise of the digital workflow in dentistry is based on three elements. The first is data acquisition, as in several scanning technologies. The second is data manipulation and processing, created using computer-aided design software. Finally, the third element is the processed data is used to fabricate structures in the desired material through computer-aided manufacturing.

Katkar, Taft and Grant (2018) indicate that vat photopolymerization (PV), more specifically stereolithography (SLA) and digital light processing (DLP) are the most popular impression types in dentistry and that these impression types require a good amount of post-processing to remove substrates, remove unused material and to complete curing of the material. In addition, the authors point out that this type of impression works based on the exposure of a light source to a vat of layered photosensitive liquid resin. Kim *et al.* (2020), point out that a factor used to optimize the post-curing of the material is to place it in a plaster mold, causing the settlement of the 3D printed prosthesis to have its polymerization contraction reduced, enabling the manufacture of more accurate prostheses.

Barazanchi *et al.*, (2017) reaffirm that the wide range of additive techniques and materials available expand the potential of 3D printing to many other applications in dentistry. Methani *et al.*, (2020) identify four different vat photopolymerization (PV) technologies that can be differentiated based on the light source employed for polymerization: stereolithography (SLA), digital light processing (DLP), liquid crystal display (LCD), also called daylight polymer printing (DPP), and continuous liquid interface production (CLIP).

Methani *et al.* (2020) describe the SLA procedure (Figure 3) as a 3D printing technique that uses a laser or ultraviolet (UV) light and a digital micro-mirror device (DMD) to direct this beam of light to fabricate the desired object, layer by layer. The build platform sequentially descends into a vat filled with photosensitive resin, followed by exposure to ultraviolet light in a pattern dictated by the cross-sectional geometry of each layer of the object. To support the object against the wiping action of the build platform and gravity, support structures are used that supplement the STL file (rapid prototyping industry standard digital data transmission file) prior to printing. The manufactured part is then subjected to post-processing procedures, which involves the removal of excess resin and support structures, followed by polymerization in an ultraviolet chamber.

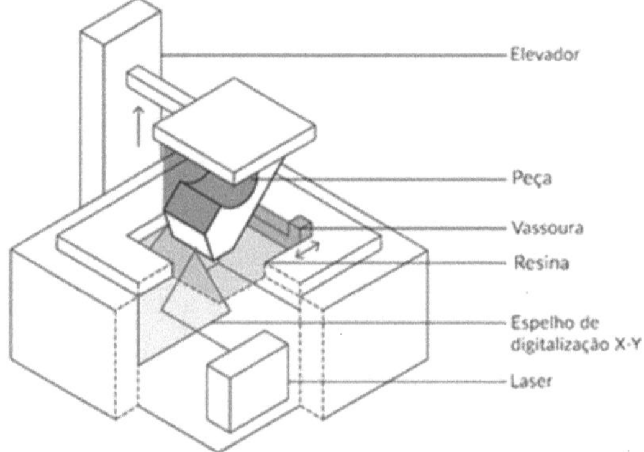

Figure 3: Operation of an SLA 3D Printer
Source: 3dlab (2020).

Elomaa *et al.* (2013) restate by indicating that stereolithography (SLA) machines use photosensitive resins as a building material, curing one layer at a time using ultraviolet or laser light and that these polymers offer much more flexibility in colour, stiffness and component modification. In addition they can also be blended with biocompatible and bioactive components as they allow uniform distribution of the added compounds and exhibit *in vitro* bioactive properties.

Digital light processing (DLP) is described by Dawood *et al.,* (2015) as a vapor polymerization technology that differs from SLA in the light source used.

Figure 4: Operation of a DLP 3D Printer
Source: 3dlab (2020).

DLP (Figure 4) uses a digital micro-mirror device (DMD) to reflect light and polymerize each layer, while LCD technology uses daylight or artificial light projected onto liquid crystal displays (LCDs) instead of UV lasers or projectors to polymerize the resin, which makes it inexpensive compared to the other technologies.

Finally, Tumbleston *et al.* (2015, *apud* METHANI *et al.*, 2020) explain that CLIP technology uses a micro mirror (DMD) to polymerize the photosensitive resin through an oxygen permeable window made of fluoropolymer that creates an area called "dead zone" where polymerization is inhibited between the window and the polymerized part.

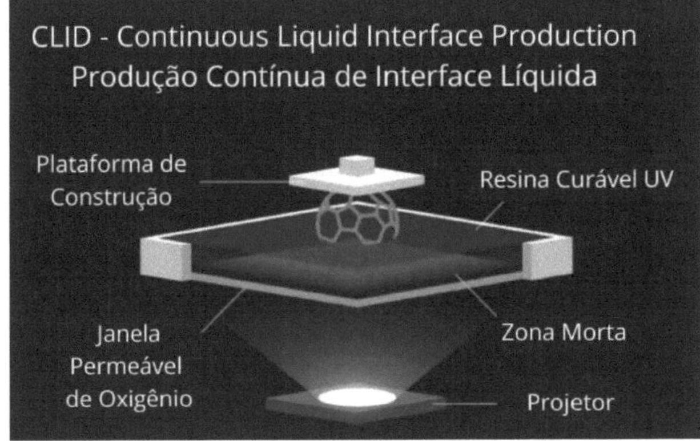

Figure 5: Operation of a CLIP 3D Printer

Source: Adapted from How it Works Team (2015).

CLIP (Figure 5) fabricates the 3D object continuously rather than gradually, thus accelerating the production process. Ultraviolet light causes photopolymerization, while oxygen inhibits it. The careful balance of this interaction of light and oxygen allows objects to be created continuously from resin. This, in addition to being fast in the process, provides better properties of the printed objects since, because they are not printed in layers, the parts are aesthetically smooth and very solid in structure.

In addition to differentiation by printing technique, according to Stansbury and Idacavage (2016), in general the resolution of a photopolymerized vat printed object is characterized by photosensitive resin properties and processing parameters, including light energy, wavelength, printer speed, build platform positioning, slicer software, printing parameters, support structures, print angulation, resin color, object geometry, and post-processing procedures.

According to Methani *et al.* (2020), vat polymerization (PV) technologies can be used to manufacture a wide range of polymer-based devices in dentistry, i.e. diagnostic casts, custom trays, positioning guides for custom abutments, dental preparation guides, temporary dental restorations, silicone indices, occlusal devices, full dentures, patterns for casting or pressing dental restorations and surgical guides. VP technologies can also be used to fabricate definitive models for muco-, dento- or implant-supported dental prostheses. However, the authors state that the dental literature regarding the accuracy of these models is insufficient.

According to Inokoshi, Kanazawa, and Minakuchi (2015), VP procedures have also been used as a tool for the fabrication of removable and full dentures. According to Chung *et al.* (2018) denture teeth produced with VP demonstrated adequate fracture resistance to be used as PT components. In addition, SLA demonstrated higher overall manufacturing accuracy than injection-molded denture bases.

In an *in vitro* study, Patzelt *et al.* (2014) compared the accuracy of milled and additive manufacturing (AM) diagnostic models obtained from digital scans performed with a laboratory scanner and three different intraoral scanners (IOS). According to the results, the SLA models showed higher accuracy than their milled counterparts, but the accuracy was still questionable at best for prosthodontic applications. According to Wu, Li and Zhang, (2017) intraoral scanners offer, unlike the traditional method, a potential to custom design a dental prosthesis and directly make a denture framework with complex patterns.

The vat light-curing (PV) technique can also be used for processing dental ceramics, where there are two additional steps: the first involves heat treatment for the removal of organic binders and the second step is high-temperature sintering for densification of the ceramic. (METHANI *et al.,* 2020)

In addition to ceramic crowns, SLA has been used to impression zirconia implants with high dimensional accuracy. However, according to Alharbi, Wismeijer and Osman, (2017) microstructural analysis of the specimens revealed substantial porosity and cracking, which implies the search for concrete evidence before the technology can transition into clinical practice. In a different direction, Lian *et al.* (2018) state that the SLA method has potential for the development of complex zirconia PPRs and other small ceramic components of complex shape that are difficult to fabricate by other techniques.

In addition to the vat light-curing (PV) method, there are 3D printers that work with material extrusion (ME), where a filament is used that is extruded using a heated extruder of known diameter. This is the technology present in most cheap desktop printers used by hobbyists. However, its application in dental and medical use is restricted to model making, which also requires some kind of post-processing. (KATKAR, TAFT and GRANT, 2018)

Methani *et al.,* (2020) corroborate by indicating that due to its low cost, ME is one of the most widely used 3D printing technologies. In general, material extrusion (ME), more commonly known as fused deposition modeling (FDM), uses thermoplastic polymers that are extruded from the printer's heated nozzle to build the 3D object layer by layer on a build platform that moves in the z-axis.

A very famous filament used by FDM 3D printers is PEEK (thermoplastic organic polymer), however in a study coordinated by Tasopoulos *et al.,* (2020) the authors obtained that the resilience and long-term functionality of PEEK clamps, as well as the biological and structural stability of the two-piece plug parts (male-female coupling), still need to be determined.

According to Barazanchi *et al.* (2020) these polymers used include a wide variety of substances, allowing their applications to diversify. However, according to Liu, Leu and Schmitt (2006), the surface finish and final product quality of additive manufacturing (AM) are inferior compared to other technologies.

Lee *et al.,* (2020) developed a new technique that encompasses both DLP and FDM to produce a new hybrid dental model for full-arch tooth models, which would enable more accurate fabrication and verification of dental prostheses. Al-Rimawi et al., (2019) also point out that DLP (digital light processing) printer may present a viable treatment option for the restoration of self-transplanted teeth.

Another 3D printing method, multi-jet (MJP), also known as polyjet (PJ), is presented by Yoo *et al.* (2021). It works by injecting photopolymer droplets and ultraviolet light subsequently polymerizes the polymer to form a 3D model. MJP is known to be more accurate than other 3D printing techniques, but is more time consuming and expensive than other fabrication methods.

According to Mazzoli (2013), another promising 3D printing technique for the dental field is powder-based fusion (PBF), which uses powdered substrates that are spread on the machine platform in layers with the help of roller blades. Three types of PBF technologies are currently available: selective laser melting (SLM), selective laser sintering (SLS) and electron beam melting (EBM). Among them, SLS uses a high-power laser beam (Na:YAG) that heats the powder in a sealed chamber to a temperature just below the melting point. This results in partial melting and melting of powder particles layer by layer. Methani *et al.* (2020) indicates that SLM is similar to SLS except that the powder is heated to its melting temperature, resulting in complete melting of the powder particles.

Early SLS machines did not use vacuum in the build process, and the laser diameter and resistance were set incorrectly to produce a dense end product. All this made the first SLS machines inefficient and complicated to use for the production of metal structures for use in load capacity. Ucar et al., (2009, apud BARAZANCHI et al., 2017) indicate that a variation of the technique, however, called direct metal laser sintering (DMLS), is promising, producing dense end products. Studies comparing the accuracy of porcelain-fused-to-metal (PFM) crowns produced with DMLS technology have found that it has a satisfactory marginal fit for use in dental prostheses.

Finally, according to Murr *et al.,* (2012) the EBM uses a beam of high-energy electrons in a noble gas environment such as argon. A tungsten filament (cathode) is then heated to a temperature of 3000°C to emit electrons where the anode causes an acceleration of the electron beam, directed towards the powder metal with the help of magnetic coils. The collision of high-energy electrons with the powdered metal transforms its kinetic energy into heat, which is responsible for the melting of the metal powder.

According to Methani *et al.,* (2020) PBF technologies can be used to fabricate cobalt-chromium (Co-Cr) and titanium (Ti) metal frameworks for PPRs and fixed dental prostheses supported by teeth or implants. Katkar, Taft and Grant (2018) complement by saying that different powdered materials such as metals, nylon and other polymers, can be used, being fused through a laser or electron beam source. The authors further reiterate indicating that this technology requires a good amount of post-processing and is used to print cobalt chrome and titanium structures for fixed and removable prostheses in dentistry.

Bae *et al.,* (2020) indicate that 3D metal-printing technologies seem reliable as an alternative to casting methods in terms of the fit of fixed dental prostheses. However, to analyze the factors influencing fabrication and confirm the results of this review, further controlled laboratory and clinical studies are needed.

According to Barazanchi *et al.,* (2017) ceramics produced using additive techniques still have problems with anisotropic shrinkage when synthesizing the raw fabricated ceramic and exhibit a step surface effect due to the nature of the fabrication process, and so far have only proven useful as a structural fabric. Hsu *et al.* (2019) state in this same vein that a zirconia dental restoration does not meet clinical levels in strength and flexural levels.

According Roseti *et al.,* (2017) it is highlighted that, despite the encouraging results, the clinical approach of Tissue Engineering has not yet occurred on a large scale, due to the need for further studies, in addition to its high manufacturing costs and the difficulty in obtaining regulatory approval. In this same line of reasoning, the authors reach the conclusion that additive manufacturing is promising and offers new possibilities in the field of dental prosthesis, although its application is still limited.

An understanding of these limitations and developments in materials science is crucial before considering additive manufacturing as an acceptable method for the fabrication of dental prostheses (ALHARBI; WISMEIJER; OSMAN, 2017).

As for metallic productions, the structures result in a poor surface finish, due to the use of a polymer to help bind the metal powder during sintering. In this way a porous structure is produced, besides which other steps involving infiltration are required to achieve sufficient density, resulting in a poor finish (BARAZANCHI *et al.,* 2020).

Sing *et al.,* (2017) confirm in this line of reasoning that 3D printing holds great promise for processing patient-specific metal implants to create complex architectures with controlled internal features at nano, micro and macro scale and customised external geometry. However, at present, there are a number of issues that need to be addressed for these sophisticated implants to be used clinically, including advances in visualization and modeling techniques.

Barazanchi *et al.,* (2017) point out that one of the most commonly used metals for substructure of dental prostheses in this method is CoCr, which has increased in popularity, partly due to the increasing cost of precious alloys and economic pressure on patients.

According to Li (2015, *apud* BARAZANCHI *et al.*, 2017) research on the properties of CoCr revealed that not only does it have an advantage over precious alloys in terms of cost, but they also rely on good bonding characteristics with porcelain, a higher Young's modulus, higher hardness, lower density and good corrosion resistance compared to other metals used in prosthetic substructures. These properties make it more load tolerant in long-term prosthetics and more stable in a long-term oral environment.

Barazanchi *et al.*, (2017) conclude by saying that the overall impression of metals produced by DMLS for use in dental prostheses when compared to current manufacturing methods, subtractive manufacturing and investment casting, are favorable. However, in the case of CoCr use in dental prostheses, most of the current studies do not withstand a rigorous critique of their methodology to allow definitive conclusions.

Several tests have been performed on the subject, since the diversity of options for making prostheses is vast. Chung *et al.* (2018) performed a test comparing 3D printed and four prefabricated PT crowns. The 3D printed resin teeth were made of methacrylate-based photopolymerized resin on an SLA printer. In the indirect tensile fracture test, the 3D-printed resin and prefabricated crowns showed similar fracture strength, with a single difference that the ones made by 3D printing did not show quasi-plastic deformation. Regarding chipping resistance, the 3D printed resin teeth had similar result to two of the prefabricated teeth and inferior result to the other two.

Tahayeri *et al.* (2018) reaffirm in another line of reasoning, pointing out that a provisional restorative material for 3D printing allows sufficient mechanical properties for intraoral use, despite the limited accuracy of the system. Eftekhar Ashtiani *et al.*, (2018) conclude in this same sense in their *in vitro* study that the conventional method for making intracoronal restorations yields more accuracy than the 3D printing method and no differences were found between the methods using the 3D printer. In a study on another topic conducted by Stoop *et al.* (2019), it is concluded that it is possible to make acceptable fitting 3D printed resin grafts in custom shapes using computer aided design and 3D printing techniques.

Dobrzański *et al.*, (2019) in their research arrive at an answer that compared to the base model, the digital model has a smaller volume than the object being mapped. Digital models based directly on the print must be resized 0.09 - 0.12% to match the dimensions of the base model, so adjustments are possible to be measured and performed with each process.

Chung *et al.*, (2018) conclude that 3D printing combined with CAD/CAM methods can automate the manufacture of PTs, besides that 3D printing technology for prostheses will not only make the base of the prosthesis, but also its artificial teeth. In this same sense, Lee, Lee and Lee, (2017) come to the conclusion that the 3D printing method is considered applicable not only in the production of provisional restorations, but also in the production of dental prostheses with a higher level of completion.

Dikova *et al.,* (2018) reach a similar result in their study, which evaluated the wear resistance of 3D printed resin material compared to milled and conventional resins. Their findings are that 3D printed resin showed no significant difference in maximum depth loss or wear volume loss compared to milled and self-curing resins. No significant difference was revealed depending on the scrapers in maximum depth loss or wear volume loss. The results suggest that 3D printing with resin materials offers adequate wear resistance for dental use. Chung *et al.* (2018) reach this same conclusion that 3D printing with resin materials offers adequate wear resistance for dental use.

In contrast, Della Bona *et al.* (2021) point out that aesthetic appearance, wear resistance and dimensional accuracy are the main current clinical limitations that restrict the progression towards the production of functional parts with 3D printing, which may explain the absence of clinical trials and reports on permanent/definitive restorative frameworks and materials. In this regard, Katkar, Taft and Grant (2018) state that accuracy depends on the type of 3D printer, material and thickness of each layer. Chia and Wu (2015), on the other hand, point out that the main problem with the use of 3D printing is the development of biocompatible materials.

Totu *et al.,* (2017) conducted a study on the production of prostheses using nanocomposites of polymethylmethacrylate (PMMA) and titanium dioxide (TiO2), providing material savings with cost and resource reduction, since the incorporation of titania in the polymeric matrix of PMMA was shown to have antibacterial effects, specifically on *Candida* species. However, mechanical and biocompatibility tests need to be performed in order to extend the manufactured prostheses for clinical use. Prasad *et al.,* (2017) confirm by pointing out that materials engineering can add significant value to future biomaterials and next generation implants can be fabricated with nanomaterials and serve various purposes.

In a similar line of reasoning, Saratti, Rocca and Krejci (2019) put forward that bioinspired concepts have already been successfully applied in a variety of engineering fields to increase the toughness and strength of engineered materials and the area of technology with the greatest potential to unlock the development of these new approaches is additive manufacturing. Thus, three-dimensional (3D) printing technologies also offer a promising new prospect for regenerating dental tissues. This may open new research avenues in dentistry that will enhance the clinical performance of engineered dental materials.

In a study by Yoo *et al.* (2021) on the dimensional accuracy of dental models for prosthetics fabricated by different 3D printing technologies, the authors concluded that polyjet (MJP) printed models showed significantly higher precision (accuracy) than DLP and SLA models, and no significant differences in accuracy were found. They concluded that overall, all 3D printed resin models produced by DLP, MJP and SLA techniques showed clinically acceptable ranges. Once in a study by Moon *et al,* (2021) 3D printing was more accurate with less deflection and shrinkage when a DLP printer was used for short/small unit restorations only.

Tahayeri *et al.,* (2018) point out that the technology to enable prosthodontic procedures is already available in the dental market, but information is currently lacking on the performance of both 3D printable dental materials and 3D printers compatible with them. Another topic raised is that existing 3D printing companies in the area have traditionally marketed printable dental materials that are compatible only with their respective printing systems. These are generally expensive (over $50,000) and of limited availability. However, the widespread use of low-cost 3D printers (below $5,000) suggests a need for better characterization of existing 3D printable dental materials with readily accessible printing systems.

In their study of fully 3D printed removable dental prostheses, Anadioti *et al.* (2020) conclude that the current limitations of 3D printing in dentistry include elimination of the initial consultation without reliable virtual esthetic evaluation, lack of retention with printed polymers requiring realignment for clinical acceptance, inability of balanced occlusion that may compromise the stability of the prosthesis or potentially influence bone resorption, and long-term colour instability leading to compromised esthetics. Shin *et al., (*2020) in their article point out that discoloration should be considered when using 3D printing resins for restorations. Another important point raised by Chang *et al.,* (2018) is that there is a challenge between balancing the biological and mechanical demands of dental prostheses made by 3D printing.

Anadioti *et al.* (2020) conclude by pointing out that the currently recommended uses for fully 3D printed prostheses are provisional or immediate prostheses as well as customized tray or registration base fabrication for conventional workflows. Well-planned clinical studies are needed to scientifically prove the claimed advantages of this technology.

CONCLUSIONS

A systematic research was conducted in search of answers to identify the main types of 3D printing techniques to make dental prostheses and present what are the main limitations of 3D printing in the area of dentistry. This reflection was initially due to the need for an organization about the existing studies in this area, since the techniques and possibilities are extremely diverse. After applying the entire research protocol (*Methodi Ordinatio*), answers began to emerge and the systematic organization helped in the presentation of the results.

The main means identified for the production of dental prostheses using 3D printing were Stereolithography (SLA), Digital Light Processing (DLP) and Direct Metal Laser Sintering (DMLS), since they are the most cited, used and studied techniques in the field. Each one has its specific characteristics, and even though they have been selected as the main ones, this does not mean that the other 3D printing techniques are unfeasible for the manufacture of dental prostheses, but that their studies and researches found nowadays are still insufficient to reach any conclusion on the subject.

The main limitations are the cost/access to technology, the development of biocompatible materials, the esthetic appearance, long-term color instability and the need for initial consultation without reliable virtual esthetic evaluation. In the case of wear resistance and dimensional accuracy, there are articles that can counteract these arguments and achieve satisfactory results in the face of this need, so they will not be classified as limitations, since they have already been overcome.

A future more specific study is required, given that techniques and limitations are diverse. Thus, conducting a research limiting it in relation to the type of technique or the limitation that needs to be overcome would help achieve deeper results, so that new solutions and technologies are presented.

Acknowledgements

This work was carried out with the support of the National Council for Scientific and Technological Development (CNPq) and thanks for all the support, both from the institution and the research laboratory and organizations and society of UTFPR.

REFERENCES

[1] MENDONÇA T. Dental mutilation: rural workers' conceptions of responsibility for tooth loss. Public Health Rep 2001; 17:1545-7.

[2] SANTOS, Flaviana. The early loss of permanent teeth and the challenges of changing this reality in a needy community. 2014.

[3] COSER M. C. Caries frequency and loss of the first permanent molars. Revista Gaúcha de Odontologia. Porto Alegre: V. 53, n. 1, p. 01-84, 2005.

[4] PINTO, V. G. Epidemiology of oral diseases in Brazil. 1997, pp. 27-41. In L. Krieger. Promoção de saúde bucal. Artes Médicas, São Paulo.

[5] PINTO, V. G. Saúde bucal coletiva. 2000. Ed. Santos, São Paulo.

[6] SILVA, M. E. DE S. E; MAGALHÃES, C. S. DE; FERREIRA, E. F. E. Dental loss and expectation of prosthetic replacement: a qualitative study. Ciência & Saúde Coletiva, v. 15, n. 3, p. 813-820, May 2010.

[7] GUIMARÃES M. M. and MARCOS B. 1996. Expectation of tooth loss in different social classes. Revista do Conselho Regional de Odontologia de Minas Gerais.

[8] MARIA DUARTE VARGAS, Andréa; HELOÍSA PAIXÃO, Helena. Tooth loss and its meaning in the quality of life of adult users of public oral health service of the Boa Vista Health Center, in Belo Horizonte. Belo Horizonte MG, 2005. Course Conclusion Paper (School of Dentistry) - Federal University of Minas Gerais.

[9] GUIMARÃES, Mirna Rodrigues Costa et al. Challenges for the provision of dental prostheses in the public health network. 2017. Dissertation (School of Dentistry) - Ufmg - Federal University of Minas Gerais.

[10] GOFFMAN E. 1988. Stigma: notas sobre a manipulação da identidade deteriorada. Editora Guanabara Koogan, Rio de Janeiro.

[11] FERREIRA, André A. A.. Pain and tooth loss: social representations of oral health care. 2005. Biofísica e Farmacologia - Universidade Federal do Rio Grande do Norte.

[12] CALEJO, A. A. A. Fabrication of total prosthesis in CAD/CAM. Core.ac.uk, 2017.

[13] LUIZA. How Much Does a Fixed Dental Prosthesis Cost (2021). Available at: <https://hospedario.com.br/quanto-custa-uma-protese-dentaria-fixa/>. Accessed on: 11 Jul. 2021.

[14] MARGARIDA VERDE PEREIRA RAMOS DA SILVA, Ana. Processing By 3d Printing Of Mandibular Full Dentures. Coimbra, 2020. Dissertation (Mechanical Engineering) - University of Coimbra.

[15] GORNI, A. A. Introduction to rapid prototyping and its processes. 2001. Available at: <http://www.gorni.eng.br/protrap.html>. Accessed on: 11 jul. 2021.

[16] YAN, Q. et al. A Review of 3D Printing Technology for Medical Applications. Engineering, v. 4, n. 5, p. 729-742, Oct. 2018.

[17] OGLIARI, Fabrício. 3D Printing in Dentistry: everything you need to know. Yller, 6 Dec 2018.

[18] DUTRA, D. M. et al. Applicability of rapid prototyping in dentistry - a literature review. Revista de Ciências Médicas e Biológicas, v. 16, n. 1, p. 89, 14 Jul. 2017.

[19] ZAHR, Paulo. See 9 types of dental prostheses and choose the best option for you. 2020. Available at: https://blog.odontocompany.com/. Accessed on: 21 abr. 2021.

[20] MASSON, Henrique. Discover 12 different types of dentures. 2019. Available at: http://www.henriquemasson.com.br/12-tipos-de-protese-dentaria/. Accessed on: 18 mar. 2021.

[21] KAIRALLA, Rogério Adib. Everything you need to know about dental prostheses. 2018. Available at: https://saude.abril.com.br/blog/cuide-da-sua-boca/o-que-saber-sobre-proteses-dentarias/. Accessed on: 2 abr. 2021.

[22] MOTTI, Verônica. Learn how a dental prosthesis is made of resin and 3D printing - Produteca. Available at: <https://www.produtecalab.com.br/protese-dentaria-em-resina-e-impressao-3d/>. Accessed on: 11 Jul. 2021.

[23] RODRIGUES Jr, CRUZ LMS, SARMANHO APS. 3D printer in the development of research with prostheses. Rev. Interinst. Bras. Ter. Ocup. Rio de Janeiro. 2018. v.2(2): 398-413

[24] SILVA J. V. L.; MAIA I. A. Development of assistive technology devices using 3D printing. Reflexões Sobre Tecnologia Assistiva. In: I Simpósio Internacional de Tecnologia Assistiva. Centro Nacional de Referência em Tecnologia Assistiva-CTI Renato Archer. Campinas, SP, 2nd version, 2014. Available at: https://www.cti.gov.br/sites/default/files//images/cnrta_livro_150715_digital_final_segunda_versao.pdf Accessed on: 15/04/2021.

[25] HULL, C. W. The Birth of 3D Printing, Research-Technology Management, 58:6, 25-30. 1986

[26] DABAGUE, L. A.. The innovation process in the 3D printers segment. 2014.

[27] CASEY, L. Prototype pronto. Packaging Digest, v. 46, n. 8, 2009, p. 54-56.

[28] OLIVAREZ, N. 3-D printers go beyond paper and ink: Mostly celebrated by hobbyists and geeks, 3-D printers may be commonplace one day. Buffalo News, 2010, p. C4.

[29] PAGANI, Regina Negri ; KOVALESKI , João Luiz ; RESENDE , Luis Mauricio Martins de . Advances in the composition of Methodi Ordinatio for systematic literature review. Utfpr, Ponta Grossa, 2018.

[30] PAGANI, R., KOVALESKI, J., and RESENDE, L. Methodi Ordinatio: a proposed methodology to select and rank relevant scientific papers encompassing the impact factor, number of citation, and year of publication. Scientometrics, 1-27, 2015. DOI:10.1007/s11192-015-1744-x

[31] KATKAR, R. A.; TAFT, R. M., GRANT, G. T.. "3D Volume Rendering and 3D Printing (Additive Manufacturing)." Dental Clinics of North America, vol. 62, no. 3, July 2018, pp. 393-402, dl.uswr.ac.ir/handle/Hannan/48862, 10.1016/j.cden.2018.03.003. Accessed 4 July 2021.

[32] BARAZANCHI, Abdullah, et al. "Additive Technology: Update on Current Materials and Applications in Dentistry." Journal of Prosthodontics, vol. 26, no. 2, 23 Feb. 2017, pp. 156-163, onlinelibrary.wiley.com/doi/abs/10.1111/jopr.12510, 10.1111/jopr.12510. Accessed 4 July 2021.

[33] PARK, S. H. et al. Digitally Created 3-Piece Additive Manufactured Index for Direct Esthetic Treatment. Journal of Prosthodontics, v. 29, n. 5, p. 436-442, 26 Mar. 2020.

[34] WILLIAMS, F. C. et al. Immediate Teeth in Fibulas: Planning and Digital Workflow With Point-of-Care 3D Printing. Journal of Oral and Maxillofacial Surgery, v. 78, n. 8, p. 1320-1327, Aug. 2020.

[35] NAGARAJAN, N. et al. Enabling personalized implant and controllable biosystem development through 3D printing. Biotechnology Advances, v. 36, n. 2, p. 521-533, Mar. 2018.

[36] SUN, J.; ZHANG, F.-Q. The Application of Rapid Prototyping in Prosthodontics. Journal of Prosthodontics, v. 21, n. 8, p. 641-644, 23 Jul. 2012.

[37] VAN NOORT, R. The future of dental devices is digital. Dental Materials, v. 28, n. 1, p. 3-12, jan. 2012.

[38] KIM, H. et al. Denture flask fabrication using fused deposition modeling three-dimensional printing. Journal of Prosthodontic Research, v. 64, n. 2, p. 231-234, Apr. 2020.

[39] METHANI, Mohammed M., et al. "Additive Manufacturing in Dentistry: Current Technologies, Clinical Applications, and Limitations." Current Oral Health Reports, vol. 7, no. 4, 3 Nov. 2020, pp. 327-334, link.springer.com/article/10.1007/s40496-020-00288-w, 10.1007/s40496-020-00288-w. Accessed 4 July 2021.

[40] "3D Printer SLA: Understand Everything About This Technology - 3D Lab." 3D Lab, 29 Jan. 2020, 3dlab.com.br/impressor-3d-sla/. Accessed 4 July 2021.

[41] ELOMAA, L. et al. Porous 3D modeled scaffolds of bioactive glass and photocrosslinkable poly(ε-caprolactone) by stereolithography. Composites Science and Technology, v. 74, p. 99-106, Jan. 2013.

[42] DAWOOD, A. et al. 3D printing in dentistry. British Dental Journal, v. 219, n. 11, p. 521-529, 11 Dec. 2015.

[43] "DLP 3D Printer: Learn How Its Process Works." 3D Lab, 11 Feb. 2020, 3dlab.co.uk/printer-3d-dlp/. Accessed 4 July 2021.

[44] HOW IT WORKS TEAM. "Carbon3D: 3D Printing Made Faster." How It Works, How It Works, 3 Aug. 2015, www.howitworksdaily.com/carbon3d-3d-printing-made-faster/. Accessed 4 July 2021.

[45] STANSBURY, J. W.; IDACAVAGE, M. J. 3D printing with polymers: Challenges among expanding options and opportunities. Dental Materials, v. 32, n. 1, p. 54-64, jan. 2016.

[46] INOKOSHI M, KANAZAWA M, MINAKUCHI S. Evaluation of a complete denture trial method applying rapid prototyping. Dent Mater J. 2012;31:40-6. 43. Bilgin MS, Erdem A, Aglarci OS, Dilber E. Fabricating complete dentures with CAD/CAM and RP technologies. J Prosthodont. 2015;24:576-9.

[47] CHUNG, Y.-J. et al. 3D Printing of Resin Material for Denture Artificial Teeth: Chipping and Indirect Tensile Fracture Resistance. Materials, v. 11, n. 10, p. 1798, 21 Sep. 2018.

[48] PATZELT, S. B. M. et al. Accuracy of computer-aided design/computer-aided manufacturing-generated dental casts based on intraoral scanner data. The Journal of the American Dental Association, v. 145, n. 11, p. 1133-1140, Nov. 2014.

[49] WU, J.; LI, Y.; ZHANG, Y. Use of intraoral scanning and 3-dimensional printing in the fabrication of a removable partial denture for a patient with limited mouth opening. The Journal of the American Dental Association, v. 148, n. 5, p. 338-341, May 2017.

[50] ALHARBI, N.; WISMEIJER, D.; OSMAN, R. Additive Manufacturing Techniques in Prosthodontics: Where Do We Currently Stand? A Critical Review. The International Journal of Prosthodontics, v. 30, n. 5, p. 474-484, Sept. 2017.

[51] LIAN, Q.; SUI, W.; WU, X.;YANG, F. and S.. Additive manufacturing of ZrO2 ceramic dental bridges by stereolithography, 2018.

[52] TASOPOULOS, T. et al. PEEK Maxillary Obturator Prosthesis Fabrication Using Intraoral Scanning, 3D Printing, and CAD/CAM. The International Journal of Prosthodontics, v. 33, n. 3, p. 333-340, May 2020.

[53] LIU, Q.; LEU, M. C.; SCHMITT, S. M. Rapid prototyping in dentistry: technology and application. The International Journal of Advanced Manufacturing Technology, v. 29, n. 3-4, p. 317-335, 17 Aug. 2005.

[54] LEE, D. et al. A Hybrid Dental Model Concept Utilizing Fused Deposition Modeling and Digital Light Processing 3D Printing. The International Journal of Prosthodontics, v. 33, n. 2, p. 229-231, Mar. 2020.

[55] AL-RIMAWI, A. et al. 3D Printed Temporary Veneer Restoring Autotransplanted Teeth in Children: Design and Concept Validation Ex Vivo. International Journal of Environmental Research and Public Health, v. 16, n. 3, p. 496, 11 Feb. 2019.

[56] YOO, SOO-YEON, et al. "Dimensional Accuracy of Dental Models for Three-Unit Prostheses Fabricated by Various 3D Printing Technologies." Materials, vol. 14, no. 6, 22 Mar. 2021, p. 1550, www.mdpi.com/1996-1944/14/6/1550, 10.3390/ma14061550. Accessed 4 July 2021.

[57] MAZZZOLI, A. Selective laser sintering in biomedical engineering. Medical & Biological Engineering & Computing, v. 51, n. 3, p. 245-256, 19 Dec. 2012.

[58] MURR, L. E. et al. Metal Fabrication by Additive Manufacturing Using Laser and Electron Beam Melting Technologies. Journal of Materials Science & Technology, v. 28, n. 1, p. 1-14, Jan. 2012.

[59] BAE, SOOHYUN, et al. "Reliability of Metal 3D Printing with Respect to the Marginal Fit of Fixed Dental Prostheses: A Systematic Review and Meta-Analysis." Materials, vol. 13, no. 21, 26 Oct. 2020, p. 4781, www.mdpi.com/1996-1944/13/21/4781, 10.3390/ma13214781. Accessed 4 July 2021.

[60] HSU, H.-J. et al. A comparison of the marginal fit and mechanical properties of a zirconia dental crown using CAM and 3DSP. Rapid Prototyping Journal, v. 25, n. 7, p. 1187-1197, Aug. 12, 2019.

[61] ROSETI, L. et al. Scaffolds for Bone Tissue Engineering: State of the art and new perspectives. Materials Science and Engineering: C, v. 78, p. 1246-1262, Sep. 2017.

[62] SING, S. L. et al. Directive selective laser sintering and melting of ceramics: a review. Rapid Prototyping Journal, v. 23, n. 3, p. 611-623, 18 Apr. 2017.

[63] TAHAYERI, A. et al. "3D Printed versus Conventionally Cured Provisional Crown and Bridge Dental Materials." Dental Materials, vol. 34, no. 2, Feb. 2018, pp. 192-200,

www.sciencedirect.com/science/article/abs/pii/S0109564117304475, 10.1016/j.dental.2017.10.003. Accessed 4 July 2021.

[64] EFTEKHAR ASHTIANI, R. et al. Comparison of dimensional accuracy of conventionally and digitally manufactured intracoronal restorations. The Journal of Prosthetic Dentistry, v. 119, n. 2, p. 233-238, Feb. 2018.

[65] STOOP, C. C. et al. Marginal and internal fit of 3D printed resin graft substitutes mimicking alveolar ridge augmentation: An in vitro pilot study. PLOS ONE, v. 14, n. 4, p. e0215092, 15 Apr. 2019.

[66] DOBRZAŃSKI, L.B., et al. "Application of Polymer Impression Masses for the Obtaining of Dental Working Models for the Stereolithographic 3D Printing." Archives of Materials Science and Engineering, vol. 1, no. 95, 1 Jan. 2019, pp. 31-40, yadda.icm.edu.pl/baztech/element/bwmeta1.element.baztech-1e174cfd-d711-41aa-843d-00e5c7106e33, 10.5604/01.3001.0013.1015. Accessed 6 July 2021.

[67] LEE, W.-S.; LEE, D.-H.; LEE, K.-B. Evaluation of internal fit of interim crown fabricated with CAD/CAM milling and 3D printing system. The Journal of Advanced Prosthodontics, v. 9, n. 4, p. 265, 2017.

[68] DIKOVA, T.D., et al. "Dimensional Accuracy and Surface Roughness of Polymeric Dental Bridges Produced by Different 3D Printing Processes." Archives of Materials Science and Engineering, vol. 2, no. 94, 3 Dec. 2018, pp. 65-75, www.infona.pl/resource/bwmeta1.element.baztech-a87b85fa-0aa8-49db-8f9b-78088ff736fe, 10.5604/01.3001.0012.8660. Accessed 4 July 2021.

[69] DELLA BONA, A., et al. "3D Printing Restorative Materials Using a Stereolithographic Technique: A Systematic Review." Dental Materials, vol. 37, no. 2, Feb. 2021, pp. 336-350, www.sciencedirect.com/science/article/abs/pii/S010956412030350X, 10.1016/j.dental.2020.11.030. Accessed 4 July 2021.

[70] CHIA, H. N.; WU, B. M. Recent advances in 3D printing of biomaterials. Journal of Biological Engineering, v. 9, n. 1, 1 mar. 2015.

[71] TOTU, E. E., et al. "Poly(Methyl Methacrylate) with TiO 2 Nanoparticles Inclusion for Stereolithographic Complete Denture Manufacturing - the Fututre in Dental Care for Elderly Edentulous Patients?" Journal of Dentistry, vol. 59, Apr. 2017, pp. 68-77,

www.sciencedirect.com/science/article/abs/pii/S0300571217300404 , 10.1016/j.jdent.2017.02.012. Accessed 4 July 2021.

[72] PRASAD, K. et al. Metallic Biomaterials: Current Challenges and Opportunities. Materials, v. 10, n. 8, p. 884, 31 July 2017.

[73] SARATTI, C. M.; ROCCA, G. T.; KREJCI, I. The potential of three-dimensional printing technologies to unlock the development of new "bio-inspired" dental materials: an overview and research roadmap. Journal of Prosthodontic Research, v. 63, n. 2, p. 131-139, Apr. 2019.

[74] MOON, W. et al. Dimensional Accuracy Evaluation of Temporary Dental Restorations with Different 3D Printing Systems. Materials, v. 14, n. 6, p. 1487, 18 mar. 2021.

[75] ANADIOTI, E., et al. "3D Printed Complete Removable Dental Prostheses: A Narrative Review." BMC Oral Health, vol. 20, no. 1, 27 Nov. 2020, bmcoralhealth.biomedcentral.com/articles/10.1186/s12903-020-01328-8, 10.1186/s12903-020-01328-8. Accessed 4 July 2021.

[76] SHIN, J.-W. et al. Evaluation of the Color Stability of 3D-Printed Crown and Bridge Materials against Various Sources of Discoloration: An In Vitro Study. Materials, v. 13, n. 23, p. 5359, 26 Nov. 2020.

[77] CHANG, J. Z.-C. et al. Augmentation of DMLS Biomimetic Dental Implants with Weight-Bearing Strut to Balance of Biologic and Mechanical Demands: From Bench to Animal. Materials, v. 12, n. 1, p. 164, Jan. 7, 2019.

APPENDIX 1

Final Result of Article Selection

Title	InOrdinatio	Classification	Objectives	Solution/Result/Conclusion
Scaffolds for Bone Tissue Engineering: State of the art and new perspectives	547,01	1	Present rapid prototyping techniques added to strategies based on bone structures for Bone Recovering/Cicatrization Engineering	As a conclusion, it is highlighted that, despite the encouraging results, the clinical approach of Bone Tissue Engineering has not yet taken place on a large scale due to the need for further studies, besides its high manufacturing costs and the difficulty in obtaining regulatory approval.
Metallic biomaterials: Current challenges and opportunities	235,00	2	To briefly explore the main representatives of metallic biomaterials along with the existing and emerging key strategies for surface and bulk modification used to improve biointegration, mechanical strength and flexibility of biomaterials, and discuss their compatibility with the concept of 3D printing.	In this review, currently used biomaterials were discussed with possible drawbacks and suggestions for improvement. It is new or modified materials that can address these challenges currently faced by existing implants and provide biomaterials that not only minimise the likelihood of medical complications, but possibly offer more realistic and aesthetically pleasing outcomes for patients. Materials engineering can add significant value to future biomaterials, and next-generation implants can be manufactured with

				nanomaterials that can serve multiple purposes.
Direct selective laser sintering and melting of ceramics: a review	232,00	3	present a review on the additive manufacturing process of ceramic materials, focusing on partial and total fusion. of ceramic powder by a high-energy laser beam without the use of binders.	3D printing holds great promise for processing patient-specific metal implants to create complex architectures with controlled internal features at the nano, micro and macro scale and customized external geometry. However, at present, there are a number of issues that need to be addressed for these sophisticated implants to be used clinically. This includes advances in visualization and modeling techniques.
3D printed versus conventionally cured provisional crown and bridge dental materials	218,01	4	To optimize the 3D printing of a dental material for provisional crown and bridge restorations using a low-cost stereolithographic 3D printer; and to compare its mechanical properties with conventionally cured provisional dental materials.	Our results suggest that a provisional 3D impression restorative material allows sufficient mechanical properties for intraoral use, despite the limited 3D impression accuracy of the impression system of choice (SLA).

Additive Technology: Update on Current Materials and Applications in Dentistry	197,00	5	This paper will review the current applications, materials, advantages and limitations of additive manufacturing. Because of its emerging use in manufacturing technique, particular focus will be given to the current research and manufacturing method of CoCr alloy.	The wide range of additive techniques and materials available allow the potential for many more applications in dentistry. The passive nature of additive techniques allows the fabrication of more sophisticated building structures without excessive force and much less non-recyclable waste when compared to subtractive manufacturing techniques. The fabrication of CoCr substructures for dental prostheses is an area of interest. CoCr has become more popular due to its cost and favorable properties; however, it is much more difficult to fabricate using investment casting and subtractive techniques due to high hardness and low ductility. The additive DMLS technique promises to circumvent these difficulties when fabricating CoCr structures; however, the properties of the produced structures do not appear to have been investigated sufficiently to draw definitive conclusions about their properties at this time. Further research is needed to examine the suitability of CoCr fabricated in DMLS for use as a substructure in PFM dental prostheses. Although additive

				manufacturing still has its limitations, the speed of development promises a great leap forward in applying the digital workflow model to more dental procedures.
Additive manufacturing of ceramics for dental applications: A review	181,01	6	The main objective of this review is to provide a detailed and comprehensive description of the work published in the last decade on MA of ceramic materials with possible applications in dentistry. The main printable materials and the most common technologies are also addressed, highlighting their advantages and main drawbacks.	There is no perfect technology for all materials / applications, capable alone to meet all the specificities and needs of each patient. Although very promising, the additive technology of dental ceramic materials remains little studied and more work is needed to make it a widespread technology in dentistry.

Poly(methyl methacrylate) with TiO2 nanoparticles inclusion for stereolitographic complete denture manufacturing - the fututre in dental care for elderly edentulous patients?	141,01	8	The aim of this study was to obtain a poly(methyl methacrylate) (PMMA)-TiO2 nanocomposite material with enhanced antibacterial characteristics, suitable for 3D printed dental prosthetic fabrication.	Significant improvements in polymer characteristics and good dispersion of TiO2 nanoparticles were observed at 0.4% by weight, therefore, it was used for stereolithographic prototyping of total prostheses.
Enabling personalized implant and controllable biosystem development through 3D printing	128,02	9	They cover recent clinical applications of 3D printing with a particular focus on implantable devices.	Using 3D printing techniques, risk-free surgical training on objects that are highly mimetic of natural organs, both mechanically and architecturally, can be used to interact with patients to achieve better understanding and compliance for diseases that require surgery. This would lead to better surgical planning and preparation based on local

				imprinting of tumors or damaged tissue that can lead to better decision making and fewer complications.
Additive manufacturing techniques in prosthodontics: Where do we currently stand? A critical review	128,00	10	The aim of this article was to critically review the current application of additive manufacturing (AM) / 3D printing techniques in dental prosthetics and to highlight the influence of various technical factors involved in different AM technologies.	Different MA technologies are applied in dental prosthetics, directly and indirectly for the fabrication of fixed metal copings, metal frameworks for removable partial dentures and plastic mock-ups and resin patterns for other conventional metal castings. Technical factors involved in different AM techniques influence the overall quality, the mechanical properties of the printed parts and the total cost and manufacturing time. MA is promising and offers new possibilities in the field of dental prosthetics, although its application is still limited. An understanding of these limitations and of developments in materials science is crucial before considering MB as an acceptable method for the fabrication of dental prostheses.

Additive manufacturing of ZrO2 ceramic dental bridges by stereolithography	112,00	13	This work aims to develop an additive manufacturing technique for complex dental bridges made of zirconia ceramics.	This article provides a potential technical method for fabricating complex zirconia dental bridges and other small complex shaped ceramic components that are difficult to make by other manufacturing techniques.
Mechanical properties of a newly additive manufactured implant material based on Ti-42Nb	112,00	14	Study of the Mechanical Properties of an Additive Recently Implant material manufactured on the basis of Ti-42Nb	The mechanical properties of the Co-Cr alloy depended on the manufacturing method. The specimens in the casting group showed high modulus of elasticity, while those in the milling group showed low yield strength and flexural strength. All three groups exceeded ISO standards for metal-ceramic bond strength. The SLM technique can be used for the fabrication of dental prostheses according to ISO 22674: 2016 and ISO 9693-1: 2012. Considering its many other advantages, the SLM method appears to have the potential to replace traditional fabrication methods.

Accuracy of 3-unit fixed dental prostheses fabricated on 3D-printed casts	105,00	16	The aim of this in vitro study was to evaluate the fit of 3-unit FDPs fabricated on 3D printed molds made by digital light processing and to investigate the clinical applicability of 3D printing	Two-way ANOVA showed significant differences between the 2 groups (3DP and CS) in marginal and internal root mean square (RMS) values (P <0.001). However, no significant difference was found in marginal RMS values (P = .762) between pontic and nonpontic sides. The 3DP showed significantly higher RMS values than the CS (P <0.001). Thus, it is necessary to further improve the accuracy of 3D printers before using 3D printed molds in dental prosthetics.
Use of intraoral scanning and 3-dimensional printing in the fabrication of a removable partial denture for a patient with limited mouth opening	105,00	17	Use of intraoral scanning and three-dimensional impression taking in the fabrication of a removable partial denture for a patient with limited mouth opening	Unlike the traditional method, this integrated system has the potential to custom design a dental prosthesis and directly make an RPD framework with complicated patterns.

3D Volume Rendering and 3D Printing (Additive Manufacturing)	104,00	18	Volume rendering is a set of techniques used to display a two-dimensional (D) projection of a discretely sampled 3D dataset. These volume-rendered images can be sectioned in any plane and rotated in space, allowing a 3D view of the craniofacial bone anatomy. The study aims to study the accuracy and precision of this technique for 3D printing	A recent study compared 3D-printed reconstructed rapid prototyping models and conventional plaster models for different degrees of dental crowding group and arch dimensions in the mild and moderate crowding groups. Statistically significant differences were found for all planes in all crowding categories except for crown height in the moderate crowding group and arch dimensions in the. It was concluded that the rapid prototyping models were not clinically comparable to conventional stone models regardless of the degree of crowding. Limitations include time and cost; accuracy depends on the type of 3D printer, material and build thickness.
3D printing restorative materials using a stereolithographic technique: a systematic review	100,01	19	To present, by means of a systematic review, a qualitative analysis of published studies on stereolithography-based 3D printing of restorative materials and its clinical applicability.	The rapid expansion of stereolithography-based 3D printing has been impressive and represents a major technological breakthrough with significant potential for disruption. Dentistry has demonstrated an incredible willingness to adapt materials, methods and workflows to this promising digital technology. However, aesthetic appearance,

				wear resistance, wet strength and dimensional accuracy are the main current clinical limitations that restrict the progression towards the production of functional 3D-printed parts, which may explain the absence of clinical trials and reports on definitive/permanent dental restoration materials and frameworks.
Dimensional accuracy of dental models for three-unit prostheses fabricated by various 3D printing technologies	100,00	21	The aim of this study was to examine the dimensional accuracy of 3D printed models that were prepared with teeth for three-unit fixed prostheses, especially at the margins and proximal contact areas.	For overall 3D analysis, MJP showed significantly higher accuracy (precision) than DLP and SLA techniques; however, there was no statistically significant difference in precision. For deviations at the molar tooth margins and distance to proximal contact, the MJP showed significantly accurate results; however, for a premolar tooth, there was no significant difference between the groups.
Dimensional accuracy evaluation of temporary dental restorations with	100,00	21	This study analyzed the accuracy of 3D printed temporary restorations of different sizes produced by digital light processing (DLP) and liquid crystal display (LCD) printers.	In the point deviation analysis, a significant difference in direction was exhibited in all the DLP printer restorations but only in some cases for the LCD printer. Within the limitations of this study, 3D printing was more accurate with less deviation and shrinkage when a DLP printer was used for short unit

different 3d printing systems				restorations.
Design of a patient-specific mandible reconstruction implant with dental prosthesis for metal 3D printing using integrated weighted topology optimization and finite element analysis	99,01	24	The aim of this study was to use a weighted topology optimization method to design a patient-specific mandibular implant for reconstruction and appearance restoration in patients with severe mandibular defects.	This study developed a design strategy with weighted topology optimization and fabrication for patient-specific implant production using metal 3D printing. The obtained implant reconstruction can provide good biomechanical performance and appearance recovery for oral rehabilitation.

Evaluation of internal fit of interim crown fabricated with CAD/CAM milling and 3D printing system	97,00	26	This study aims to evaluate the internal fit of the crown fabricated by CAD / CAM milling method and 3D impression method	The marginal and internal fit of the provisional restoration has a more remarkable 3D printing method than the CAD/CAM milling method. Therefore, the 3D printing method is considered applicable for not only in the production of provisional restorations, but also in the production of dental prosthetics with a higher level of completion.
Effects of printing parameters on the fit of implant-supported 3D printing resin prosthetics	96,00	29	The aim of the study was to investigate the influence of 3D printing parameters on the fit and internal gap of 3D printed resin dental prostheses.	The 3D printed prosthesis showed less internal clearance than the milled prosthesis. The marginal fit of the 3D printed resin prosthesis was clinically acceptable, and the 45o and 60o build-up orientation would be recommended when considering fit and internal clearance.

Position Accuracy of Implant Analogs on 3D Printed Polymer versus Conventional Dental Stone Casts Measured Using a Coordinate Measuring Machine	96,00	31	To compare the accuracy of implant analog positions in edentulous complete maxillary casts made of dental stone or additive manufactured polymers using a coordinate measuring machine (CMM).	Compared with CMM, the average distortion (μm) ranged from 22.7 to 74.9; 23.4 to 49.1; and 11.0 to 85.8 in the x-, y- and z-axes, respectively. CDS method (x-axis: 37.1; z-axis: 27.62) showed a significant difference compared with DLP in the x-axis (22.7) ($p = 0.037$) and for MJP1 in the z-axis (11.0) ($p = 0.003$). Regardless of the cast system, x-axis showed more distortion (42.6) compared to y- (34.6) and z-axis (35.97). Among the additive manufacturing technologies, MJP2 showed the highest (64.3 ± 83.6), and MJP1 (21.57 ± 16.3) and DLP (27.07 ± 20.23) the lowest distortion, which was not significantly different from CDS (32.3 ± 22.73) ($p > 0.05$).
3D printing of resin material for denture artificial teeth: Chipping and indirect tensile fracture	95,00	32	This study evaluated chipping and indirect tensile fracture strength of 3D printed resin material (Dentca 3D printed resin denture) compared to conventionally prefabricated resin denture teeth (Premium-8, Overcoming, SR-Orthosit-PE and Preference).	The 3D printed resin teeth exhibited vertical fracture of the loaded cuspid without deformation in chipping. The 3D printed resin teeth showed simultaneous fracture of two cusps in indirect tensile fracture, unlike other teeth. The results of this study suggest that 3D printing technology using resin materials provides adequate fracture resistance for use in artificial dentures.

resistance				
Digitally Created 3-Piece Additive Manufactured Index for Direct Esthetic Treatment	95,00	33	This article describes a digital workflow protocol for treatment planning and esthetic rehabilitation with direct composite restorations.	This procedure facilitated treatment planning procedures as well as aided direct composite restorative procedures by providing several advantages compared to conventional procedures, such as accurate translation of the digital diagnostic wax-up in the patient's mouth, horizontal path of silicone index insertion, and minimized clinical intervention time.
A hybrid dental model concept utilizing fused deposition modeling and digital light processing 3d printing	94,00	35	Combining the use of DLP 3D printing and fused deposition modeling (FDM). Due to the characteristics of the material, the FDM accuracy of the full-arch dental model should be higher than that of DLP; however, due to the surface roughness of the FDM method, the crown preparation matrix for dental prosthesis is sometimes not accurate.	Today, many dentists use oral scanners to transfer oral information to an STL file and send it to the lab. Although technicians can fabricate a dental prosthesis without using dental models, a dental model output is essential to verify the correct occlusion and to fabricate complicated prostheses. Therefore, the demand for accurate dental models will continue to increase in the future.

			Therefore, a new hybrid dental model that combines FDM (for the full dental model) and DLP (for the specific matrix) is proposed.	In this technique, a new hybrid dental model using FDM and DLP 3D printing for full-arch tooth models was introduced. This technique would enable more accurate fabrication and verification of dental prostheses.
Three-dimensional evaluation on accuracy of conventional and milled gypsum models and 3D printed photopolymer models	93,00	36	The aim of this study was to evaluate the accuracy of dental models fabricated by conventional, milling and three-dimensional (3D) impression methods.	For full arch, the RMS value of trueness and accuracy in CON was significantly lower than in the other groups ($p < 0.05 / 6 = 0.008$), and there was no significant difference between S3P and D3P ($p > 0.05 / 6 = 0.008$) On the other hand, the RMS value of trueness in CON was significantly lower than in the other groups for all prepared teeth ($p < 0.05 / 6 = 0.008$), and there was no significant difference between MIL and S3P ($p > 0.05 / 6 = 0.008$). In conclusion, conventional plaster models showed better accuracy than digitally milled and 3D printed models.

Evaluation of the color stability of 3d-printed crown and bridge materials against various sources of discoloration: An in vitro study	92,00	37	Thus, this study aimed to evaluate the discoloration resistance and color stability of CAD / CAM blocks and 3D impression materials by assessing the color changes after exposure to food stains. The null hypothesis is that there is no difference in the discoloration characteristics due to the restoration material used, discoloration as a function of the type of dye used or storage time of the materials in the dyes.	The 3D impression resins showed color differences above the clinical limit (2.25) after storage for 7 days or longer in all experimental groups. Curry was the most prominent colorant and discoloration increased in almost all groups with increasing storage duration. This study suggests that discoloration should be considered when using 3D impression resins for restorations
Mechanical properties and metal-ceramic bond strength of co-cr alloy manufactured by selective laser melting	92,00	37	Cobalt-chromium metal (Co-Cr) is one of the biomaterials widely used in the fabrication of dental prostheses. The aim of this study was to investigate whether there are differences in metal properties and bond strength with ceramics depending on the manufacturing methods of the Co-Cr alloy.	The SLM group showed finer homogeneous crystalline microstructure and a layered structure was observed on the fractured surface. After testing the bond strength to ceramics, all groups showed a mixed failure pattern. The cast group showed the highest bond strengths, whereas there was no significant difference between the other two groups. However, all groups met the standard of bond strength according to the international standards organization (ISO) with the appropriate pass rate. The

				results of this study indicate that the SLM fabrication method may have the potential to replace traditional fabrication techniques for dental prostheses.
Reliability of metal 3d printing with respect to the marginal fit of fixed dental prosthese s: A systematic review and meta-analysis	92,00	37	This systematic review and meta-analysis aimed to evaluate the marginal fit reliability of 3D printed cobalt-chromium-based fixed dental prostheses compared to conventional casting methods.	3D Metal impression technologies appear reliable as an alternative to casting methods in terms of the fit of fixed dental prostheses. To analyze the factors influencing fabrication and confirm the results of this review, further controlled laboratory and clinical studies are needed.

Immediate Teeth in Fibulas: Planning and Digital Workflow With Point-of-Care 3D Printing	92,00	40	Three-dimensional (3D) printing at the point of care has become more common in recent years because many hospitals have established 3D printing labs. Traditional techniques to fabricate an immediate dental prosthesis for fibula and implant reconstructions have involved outsourcing to dental laboratories. This results in delays, making it suitable only for benign diseases. In the present report, we demonstrate a technique for in-house creation of a 3D printed dental prosthesis for implant placement in maxillofacial free fibula reconstruction. Our digital method reduced costs and shortened the interval between surgery compared to traditional laboratory techniques.	All 12 patients received an immediate implant-retained fixed prosthesis in fibula reconstruction. The time required to generate the 3D printed prostheses in-house was significantly less than that required to create the prostheses fabricated in a dental laboratory. Costs were also lower with 3D printed prostheses compared to dental laboratory fabricated prostheses. The digital workflow we present eliminates the delay in creating a provisional dental prosthesis fabricated in a dental laboratory for fibula and implant reconstruction. This enables immediate dental restoration for patients with malignancy previously deemed unsuitable due to the inherent delay required to use an off-site dental laboratory. A reduction in cost to create 3D printed prostheses in-house was observed compared to prostheses fabricated by a dental laboratory. Case selection is critical to predict soft tissue requirements for composite defects.

Cytocompatibility of 3D printed dental materials for temporary restorations on fibroblasts	92,00	41	This study aims to investigate the influence of dental materials manufactured by both 3D printing and self-healing technology on fibroblasts.	CLSM and SEM cell imaging revealed that the 3D printed material group showed better cell adhesion with well distributed filopodia compared to the conventional resin material group. Cell proliferation was significantly higher in the 3D printed materials. The superior cytocompatibility of the specimens fabricated by 3D printing and polishing process was demonstrated with the evidence of better cell adhesion and higher cell proliferation.
Antibacterial drug-releasing polydimethylsiloxane coating for 3d-printing dental polymer: Surface alterations and antimicrobial effects	91,01	42	This study evaluated the effects of a polydimethylsiloxane (PDMS)-based coating material loaded with chlorhexidine (CHX) on the surface microstructure, surface wettability, and antibacterial activity of 3D-printed dental polymer.	First, mesoporous silica nanoparticles (MSN) were used to encapsulate CHX, and a combination was added to PDMS to synthesize an antibacterial agent-releasing coating substance. Then, a thin coating film was formed on the 3D printing polymer using oxygen plasma and heat treatment.

Influence of the Postcuring Process on Dimensional Accuracy and Seating of 3D-Printed Polymeric Fixed Prostheses	91,00	44	The aim of this study was to evaluate the effects of different postcuring methods on the fit and dimensional accuracy of 3D printed full-arch polymeric fixed prostheses. A study plaster model with four prosthetic implant abutments was prepared.	In general, the mean marginal and internal gaps of the cured prostheses were the smallest when the printed prostheses were cured with seating on the plaster model (P <0.05). Regarding the accuracy of fit, the presence of supports during the postcuring process did not make a significant difference. The error in intermolar distance was significantly lower in the model seating condition than in the other conditions (P <0: 001). Seating the 3D printed prosthesis on the plaster model reduces adverse deformation in the postcuring process, enabling the fabrication of prostheses with favorable fit.
Augmentation of DMLS biomimetic dental implants with weight-bearing strut to balance of biologic and mechanical	91,00	45	The aim of this study was to optimize the design of Ti6Al4V dental implants fabricated by direct metal laser sintering (DMLS), minimize elastic mismatch, allow maximum bone growth, and improve long-term implant fixation.	The results showed that increasing porosity decreased mechanical properties, while increasing with a longitudinal connecting rod can improve mechanical strength. Maximum expression of the alkaline phosphatase gene of MG63 cells achieved in Ti6Al4V discs with 60% porosity. In vivo experiments showed good incorporation of bone into the porous structures of DMLS dental implant, resulting in higher pull-out strength. In summary, we introduced a new

demands: From bench to animal				design concept by augmenting the implant with a longitudinal weight-bearing strut to achieve the optimal combination of high strength and low modulus of elasticity; our results showed that there is a chance to achieve a balance between biological and mechanical demands.
3D printed complete removable dental prostheses: a narrative review	91,00	46	The aim of this article is to review the available literature on three-dimensionally printed total prostheses in terms of new biomaterials, manufacturing techniques and workflow, clinical performance and patient satisfaction.	Milled dentures have been studied more than 3D printed dentures in the currently available literature. The limited number of clinical studies, mainly case reports, suggests that the current indications for 3D printing in the denture fabrication process are custom trays, registration bases, temporary or immediate dentures, but not the fabrication of permanent dentures. Limitations include poor esthetics and retention, inability to balance the occlusion, and low printer resolution. Initial studies on digital dentures have shown promising short-term clinical performance, positive patient-related outcomes and reasonable cost-effectiveness. 3D printing has the potential to modernize and streamline prosthetic fabrication

				techniques, materials and workflows. However, more research is needed on existing and developing materials and printers to enable advancement and increase its application in removable dentures.
Digital fabrication of removable partial dentures made of titanium alloy and zirconium silicate micro-ceramic using a combination of additive and subtractive manufactu	90,00	47	This work aims to present a new design for removable partial dentures (RPDs) for partially edentulous patients to improve the efficiency and quality of RPD fabrication. Additive and subtractive fabrication technologies and the bonding of zirconium silicate microceramics in the esthetic zone are used here.	The RPDs were conventionally evaluated by a physician, were found to be accurate and satisfactory to both patient and physician. This new method provides a way to fabricate RPDs with a combination of additive and subtractive manufacturing technologies. The framework design was different from a conventional framework because it contained the crown retainers and the traditional base retainer no longer existed. Ceramage bonding was used to replicate the gingival anatomy in the esthetic zone. The new RPDs provided precision and were less time consuming to produce than those produced

ring technologies				with the traditional method. The new method allows digital fabrication of almost all RPDs.
Flexural strength of 3D-printing resin materials for provisional fixed dental prostheses	90,00	48	This study compared the flexural strength of 3D printed three-unit fixed dental prostheses with that of conventionally fabricated and milled restorations.	The DLP and SLA groups presented significantly higher flexural strength than the conventional group ($p < 0.001$). No significant difference was observed in flexural strength between the DLP and SLA groups. The FDM group showed only teeth but no fracture. The results of this study suggest that provisional restorations fabricated by DLP and SLA technologies provide adequate flexural strength for dental use.
Additive Manufacturing in Dentistry: Current Technolog	90,00	49	This review aimed to illustrate the utility of additive manufacturing technologies for the fabrication of polymer, metal and ceramic components within the	Vat polymerization, material jetting and powder-based fusion technologies have existing clinical applications, primarily using polymers and metals. Additive manufacturing technologies need further

ies, Clinical Applications, and Limitations			limits of their current and potential clinical applications in dentistry.	development for use with ceramic materials for dental applications.
PEEK maxillary obturator prosthesis fabrication using intraoral scanning, 3D printing, and CAD/CAM	90,00	51	This report presents a digital technique for fabricating a maxillary obturator prosthesis with a two-piece hollow bulb.	Intraoral scanning, 3D printing and CAD/CAM were successfully used for the fabrication of a two-piece PEEK maxillary RDP obturator. The resilience and long-term functionality of the PEEK staples as well as the biological and structural stability of the two-piece obturator parts (male-female coupling) still need to be determined.
Comparison of dimensional accuracy of conventionally and digitally manufactured intracoronal restoration	89,00	52	The aim of this in vitro study was to compare the dimensional accuracy of intracoronal restorations fabricated with digital and conventional techniques.	Within the limitations of this in vitro study, the conventional method yielded more accuracy than the 3D printing method, and no differences were found between the methods that used the 3D printer (CP and IP groups).

s				
Wear resistance of 3D printing resin material opposing zirconia and metal antagonists	89,00	53	This study evaluated the wear resistance of 3D printing resin material compared to conventional milling and resin materials.	No significant difference was revealed depending on the scrapers in maximum depth loss or wear volume loss. In SEM views, the 3D printed resin showed cracking and separation of interlayer bonds when opposing the metallic abrasive. The results suggest that 3D printing with resin materials offers adequate wear resistance for dental use.
The potential of three-dimensional printing technologies to unlock the development of new 'bio-inspired' dental materials: an	88,00	54	Bioinspiration is an approach in engineering that aims to optimize artificial systems by borrowing biological concepts from nature. This review aims to summarize the fundamental aspects employed by nature to prevent premature dental failure. Based on these findings, it then defines and evaluates rules for "postmodern" manufacturing processes	Bioinspired concepts have already been successfully applied in a variety of engineering fields to increase the toughness and strength of engineered materials. The area of technology with the greatest potential to unlock the development of these new approaches is additive manufacturing. Consequently, these technologies and concepts can be applied to dentistry to improve the mechanical properties of dental restorations. Three-dimensional (3D) printing

overview and research roadmap			to mimic or regenerate complex biological systems.	technologies also offer a promising new perspective for regenerating dental tissues. Considering the limitations of conventional and subtractive computer-aided design/computer-aided manufacturing (CAD/CAM) methods, future research should focus on new additive 3D printing techniques. This may open up new research avenues in dentistry that will enhance the clinical performance of artificial dental materials.
Comparison of student's perceptions between 3D printed models versus series models in paediatric dentistry hands-on session	87,00	55	The aim of this study was to develop and evaluate a 3D printed model for training in pediatric dentistry and compare it with the reference model used in our faculty.	There are still many ways to improve these models. For example, modifying the quality of the resins can improve the milling feel and the design can be improved to achieve better contact points. However, these 3D models offer the possibility of giving the patient a more central place in the education of future professionals.

Dimensional accuracy and surface roughness of polymeric dental bridges produced by different 3D printing processes	87,00	56	This study evaluated the wear resistance of 3D printing resin material compared to conventional milling and resin materials.	The 3D printed resin showed no significant difference in maximum depth loss or wear volume loss compared to the milled and self-curing resins. No significant difference was revealed depending on the scrapers in maximum depth loss or wear volume loss. In SEM views, the 3D printed resin showed cracking and separation of interlayer bonds when opposing the metal abrasive. The results suggest that 3D printing with resin materials offers adequate wear resistance for dental use.
Non-Ionized, High-Resolution Measurement of Internal and Marginal Discrepancies of Dental Prostheses Using Optical Coherence Tomograp	86,00	57	In this paper, we proposed the use of swept-source optical coherence tomography (OCT) as a new application for non-ionized, high-resolution measurements of internal and marginal discrepancies at four anatomically critical points, such as occlusal, angle, axial, and margin during denture fixation	The demonstrated qualitative and quantitative evaluations can be well used for the assessment of the internal and marginal fit of teeth

hy				
3D printed temporary veneer restoring autotransplanted teeth in children: Design and concept validation ex vivo	85,00	58	The aim of this study was to develop and validate a digital solution for temporary restoration of autotransplanted teeth by means of 3D printing.	The present concept of using temporary veneers that are designed and fabricated with CAD / CAM (computer aided design / computer aided manufacturing) technology using a DLP (digital light processing) printer may present a viable treatment option for the restoration of autotransplanted teeth.
Marginal and internal fit of 3D printed resin graft substitutes mimicking alveolar ridge	82,01	60	The aim of this in vitro study was to evaluate the marginal and internal fit of 3D printed resin grafts as they can be used for alveolar ridge augmentation.	Within the limitations of the study, it could be demonstrated that it is possible to fabricate 3D printed resin grafts with acceptable fit into custom shapes by combining CBCT scans and computer-aided design and 3D printing techniques.

augmentation: An in vitro pilot study				
Dimensional capability of selected 3DP technologies	82,00	61	This paper aims to report the results of an investigation to compare three different three-dimensional printing (3DP) or additive manufacturing technologies [i.e., fused deposition modeling (FDM), stereolithography (SLA), and material jet (MJ)] and four different pieces of equipment (FDM, SLA, MJP 2600, and Object 260) in terms of dimensional process capability (dimensional accuracy and surface roughness). It provides a comprehensive and comparative understanding on the level of achievable dimensional accuracy, repeatability and surface roughness of commonly used 3DP technologies. It is hoped that these	The results show that FDM technology with the equipment used results in a rough surface and low dimensional accuracy. The SLA printer produced a smoother surface, but resulted in distortion of thin features (<1 mm). The MJ printers, on the other hand, produced comparable surface roughness and dimensional accuracy. However, the ProJet MJP 3600 produced sharper edges compared to the Objet 260 which produced rounded edges. This paper, for the first time, provides a comprehensive comparison of three different commonly used 3DP technologies in terms of their dimensional capability and surface roughness without post-processing. Therefore, it provides reliable guidance for design consideration and printer selection based on the target application.

			findings will assist other researchers and industrialists in choosing the right technology and equipment for a given 3DP application.	
Application of polymer impression masses for the obtaining of dental working models for the stereolithographic 3d printing	82,00	62	The aim of the work is to perform measurements of digital dental models obtained by scanning prosthetic impressions using engineering CAD software and find dimensional differences and scale factor for accurate reproduction of patient's dental dimensions.	It was found, that compared to the base model, the digital model has a smaller volume than the object being mapped, the digital models based directly on the print must be resized 0.09 - 0.12% to match the dimensions of the base model.

A comparison of the marginal fit and mechanical properties of a zirconia dental crown using CAM and 3DSP	81,00	63	This study aims to compare the marginal fit, flexural strength, and hardness of a ceramic premolar constructed by computer-aided dental machining (CAM) and three-dimensional slurry impression (3DSP).	This study verifies whether 3DSP can be used to fabricate a zirconia dental restoration device that is as good as the one produced with the CAM dental system and that has a marginal gap smaller than the threshold value. The resulting premolar restoration devices produced by sintering the green bodies produced with 3DSP and dental CAM under the same conditions have a similar hardness value, four times that of enamel. The flexural strength of 3DSP does not meet the requirements for clinical use.
Effects of titanium dioxide and tartrazine lake on Z-axis resolution and physical properties of resins printed by visible-light 3D printers	75,00	65	The purpose of this paper is to examine the effects of the light control system that combined high refraction particles (n-TiO2 [titanium dioxide - TiO2]) and tartrazine lake dye (TL dye) on the thickness, flexural strength, flexural modulus and surface details of the 3D printed resin.	The polymerization thickness of the samples containing n-TiO2 and TL dye was lower compared to the samples with TL dye only. Samples containing more n-TiO2 and more TL dye exhibited lower flexural strength and modulus. Ramp models showed that for samples containing 1 per Per cent TL dye, when its n-TiO2 content increased from 1 to 5 per cent, the laminated surface structures became sharper. However, when the TL dye the content doubled to 2 percent, the laminated surface structures were undefined compared with 1 percent of

				counterparts containing TL dye. In visible-light 3DP, the light control system in cooperative dye with high-refraction particles provides better energy distribution and scattering control. High-refraction particles, dyes and light exposure time influenced the surface resolution and mechanical properties of 3DP products.

Claudia Tania Pincinin

Pedro Ienne Fernandes

Printed by Books on Demand GmbH, Norderstedt / Germany